"Never underestimate the power
you have to take your life
in a new direction"

-Germany Kent

I Am Fearless, Selfless and Destined for Greatness
SELF-CONFIDENCE

By Certified Life Coach and Children's Author Reea Rodney

SELF-CONFIDENCE WORKBOOK

Copyright © 2017 Dara Wisdom and Empowerment Coaching / Reea Rodney

All rights reserved. This book or any part of it, may not be reproduced, in any form without written permission.

Printed in the United States of America
ISBN: 978-0-9975059-6-2

Written by Reea Rodney
Illustrated by Alexandra Gold
Designed by FindlayCreative.com

Dara Wisdom and Empowerment Coaching

I Am Fearless, Selfless and Destined for Greatness
Improving Your Self-Confidence Workshop

Table of Contents

Page 4	Introduction
Page 5	About the Author
Page 6	What is Self-Confidence?
Page 7	Why is Self-Confidence Important?
Page 8	What are the benefits of Being Confident?
Page 9	How Can we Teach our Children to be More Self-Confident?
Page 10	Fun Ways to Improve Your Self-Confidence At Home And School
Page 11	Parent Tips for Helping your Child Promote Self-Confidence

Self-Confidence Activities

Page 12	Activity 1 - Don't Wait Too Long!
Page 13	Activity 2 - Pride Line
Page 14	Activity 3 – My Review of Me
Page 15	Activity 4 – My Personal Shield
Page 16	Activity 5 – Who Decides?
Page 17	Activity 6 – A Letter to Myself
Page 18	Activity 7 – Word Finder
Page 19	Activity 8 – Fill in The Blank
Page 20	Activity 9 – Expressions All About Me
Page 21	Activity 10 – Practice What You're Good At
Page 22	Activity 11 – Compliment yourself Go for it!
Page 23	Activity 12 – Change the Negative Talk into Positive Talk
Page 24	More by Dara Publishing and Strictly Essential Clothing

Copyright © 2017 Dara Wisdom and Empowerment Coaching / Reea Rodney

I Am Fearless, Selfless and Destined for Greatness
Improving Your Self-Confidence
Introduction

It is very easy to feel comfortable with ourselves when our accomplishments are acknowledged by a peer, family, or when we get the grades we were hoping for. But what happens if we don't meet our own expectations or the expectations of others? Or, what if something unexpected happens? Often, these unforeseen circumstances can throw us off; we may begin questioning our self-worth and doubting our abilities.

This workbook series was designed as a guide to provide children with the proper tool needed to enhance, improve and develop positive self-esteem traits and practices. Children are introduced to a self-esteem and self-evaluation activities which are created to help them recognize their strength and weakness and provide the foundation for them to build on.

About the Coach / Author
Reea Rodney

Reea Rodney is a wife and mother of three wonderful children who resides in Brooklyn, New York. Originally from Trinidad & Tobago, a small twin island located in the West Indies, she migrated to the United States in 2006 in pursuit of a better life for her family. In addition, Reea is also an Empowerment Life Coach, Children's Author, Motivational Speaker, a Childcare Provider and a Medical Assistant.

Because of her innate passion and desire to help children, Reea was inspired to write children's books via her publishing company, Dara Publishing LLC. She wanted to assist not only the children who were under her care, but children all over the world. Fueled by this purpose, Reea became a Certified Life Coach. The result? Dara Wisdom and Empowerment Coaching. In addition, Reea aspires to be a positive voice of empowerment for children that she herself lacked when she was a child.

She seeks to educate parents and young children through her dynamic mini workshops and self-improvement workbooks. Topics such as Self-Esteem, Self-Love, Self-Celebration, Self-Confidence and Bullying are topics that Reea addresses through her programs. While most of these life skills are not taught in schools they are valuable to a child's overall wellbeing and development.

I Am Fearless, Selfless and Destined for Greatness

Improving Your Self-Confidence
What is Self-Confidence?

Self-confidence is a little different from self-esteem. If you respect, appreciate and admire yourself, then you have a healthy self-esteem. If you have high self-confidence that means that you trust yourself and your abilities to get anything done.

Being self-confident is like having faith in yourself. It allows you to believe that you will achieve what you set your mind to. Self-confidence is extremely important in reaching your goals, making new friends, trying new things, speaking out loud, etc.

Now that I've shared with you what Self-Confidence is, do you think you have confidence in yourself? If the answer is Yes, please share, but if your answer is No, please write down why you don't feel confident in yourself.

I Am Fearless, Selfless and Destined for Greatness
Improving Your Self-Confidence
Why is Self-Confidence Important?

When you have confidence in yourself you are in control of the way you feel, think, act and speak. Being confident allows you to be assertive and positive.

Here are more reasons why you need to be self-confident:

Self-confidence encourages you to take risks, make mistakes, and learn from them;

When you are confident in yourself you can make decisions and stick to them, which earns you respect and keeps you accountable;

You feel more powerful because you believe in yourself;

Self-confidence will allow you to meet new children and make more friends;

You have more energy, which allows you to do more things every day.

Can you think of some great benefits you can get from being Self-Confident?

_____

_____

_____

_____

_____

I Am Fearless, Selfless and Destined for Greatness
Improving Your Self-Confidence
What is Self-Talk and Why is it Important?

Self-Talk is the way you talk to yourself about yourself. It can be done in a positive or a negative way. Here are examples:

Positive self-talk - "I am really good at doing math."

Negative self-talk - I'm not smart at doing math."

When you talk to yourself in a positive way, it can make a huge difference in the way you feel. Positive self-talk can also help you build your confidence.

When you learn to speak about yourself positively, you are more likely to keep trying even when things don't work out the way you expected it to the first time. You may also want to try new things that you weren't sure of in the past.

Now that I've explained positive self-talk can you write (5) positive things that you can say to yourself.

I Am Fearless, Selfless and Destined for Greatness
Improving Your Self-Confidence
Fun Ways to Improve Your Self-Confidence at Home and at School

There are a number of ways you can improve your self-confidence at home and at school.

- Make a new friend each day at school by introducing yourself.
- Join a drama or dance class.
- Learn to play an instrument of your choice.
- Doing role play games with family members and friends.
- Having friendly debates with parents or siblings

What fun ways can you think of to help improve your Self-Confidence at home or school?

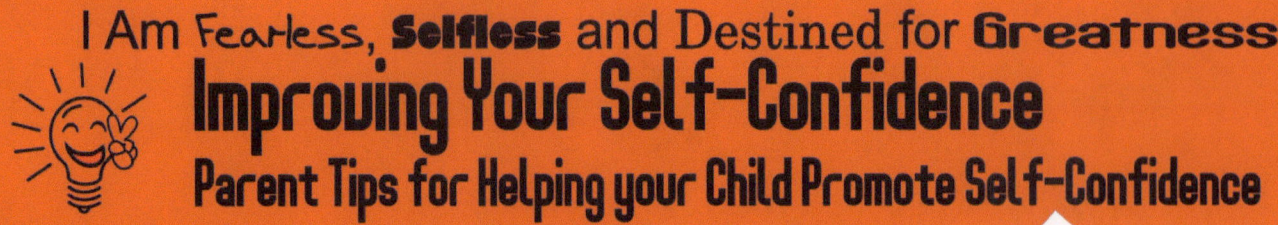

I Am Fearless, Selfless and Destined for Greatness
Improving Your Self-Confidence
Parent Tips for Helping your Child Promote Self-Confidence

Here are a few tips for parents to help your child promote self-confidence.

- Always embrace any ideas that your child may have no matter how bizarre it may be. Laughing at their ideas is never encouraging as it weakens their confidence.

- Celebrate any success or accomplishments your child may have. Children react favorably to praises for their efforts, which as a result, boost their confidence.

- Embrace your child's creativity by allowing them to express themselves in a positive manner.

- Portray confidence in your own actions as children see their parents as role models and often strive to be like them. Where there is a confident parent, it results in a confident child.

- Allow your children to experience failure without interfering. It might be hard and frustrating but children often learn more from trial and error. Remember it's not what you do for your child, but what you teach them to do for themselves to be successful for life.

Activity 1
Don't Wait Too Long!

Once you have decided that there are some things about yourself that you would like to change, don't wait too long to start changing them! You may miss some golden opportunities because you aren't prepared. Instead, get started!

In one column below, list some changes that you want to make.
Some suggestions are:
- Your habits
- Your goals
- Your grades
- Your appearance

Changes I Want to Make:	Steps I Can Take to Make these Changes:

Activity 2
Pride Line

Pride is related to self-concept. People enjoy expressing pride in something they have done that might have gone unrecognized otherwise; however it is sometimes difficult for people to say, "I'm proud that I...."

Can you to make a statement about a specific area of behavior, beginning with "I'm proud that I....". For example, you might say,
"I'm proud that I am able to read all by myself."

Below are some suggested topics for you to use in this exercise:

1. Things you've done for your parent(s)

2. Things you've done for a friend

3. Work in school

4. How you spend your time outside of school

5. Something you do often

6. Something you have shared

7. Something you are good at

8. Something you can make

Activity 3
My View of Me

This activity will give you an idea as to how you see yourself. Please rate yourself honestly, and see what you come up with "I'm so excited!"

Answer the statements below by ranking how much you agree with each one.

	Really Disagree		I kind of Agree		Really Disagree
I am smart	1	2	3	4	5
I am helpful	1	2	3	4	5
I am kind	1	2	3	4	5
I am happy	1	2	3	4	5
I am funny	1	2	3	4	5
I am supportive	1	2	3	4	5
I am talented	1	2	3	4	5
I am a good child	1	2	3	4	5
I am a good listener	1	2	3	4	5
I am good sharer	1	2	3	4	5
I am friendly	1	2	3	4	5
I am likable	1	2	3	4	5
I am awesome	1	2	3	4	5
I am brave	1	2	3	4	5
I am unique	1	2	3	4	5
I am attractive	1	2	3	4	5
I am important	1	2	3	4	5
I belong	1	2	3	4	5

Copyright © 2017 Dara Wisdom and Empowerment Coaching / Reea Rodney

Activity 4
My Personal Shield

Draw a symbol in each block of the shield to correspond with the topics listed at the bottom of the page.

1 Something I do well
2 The best compliment I have received
3 My greatest character strength
4 My favorite place
5 Something I would like to do

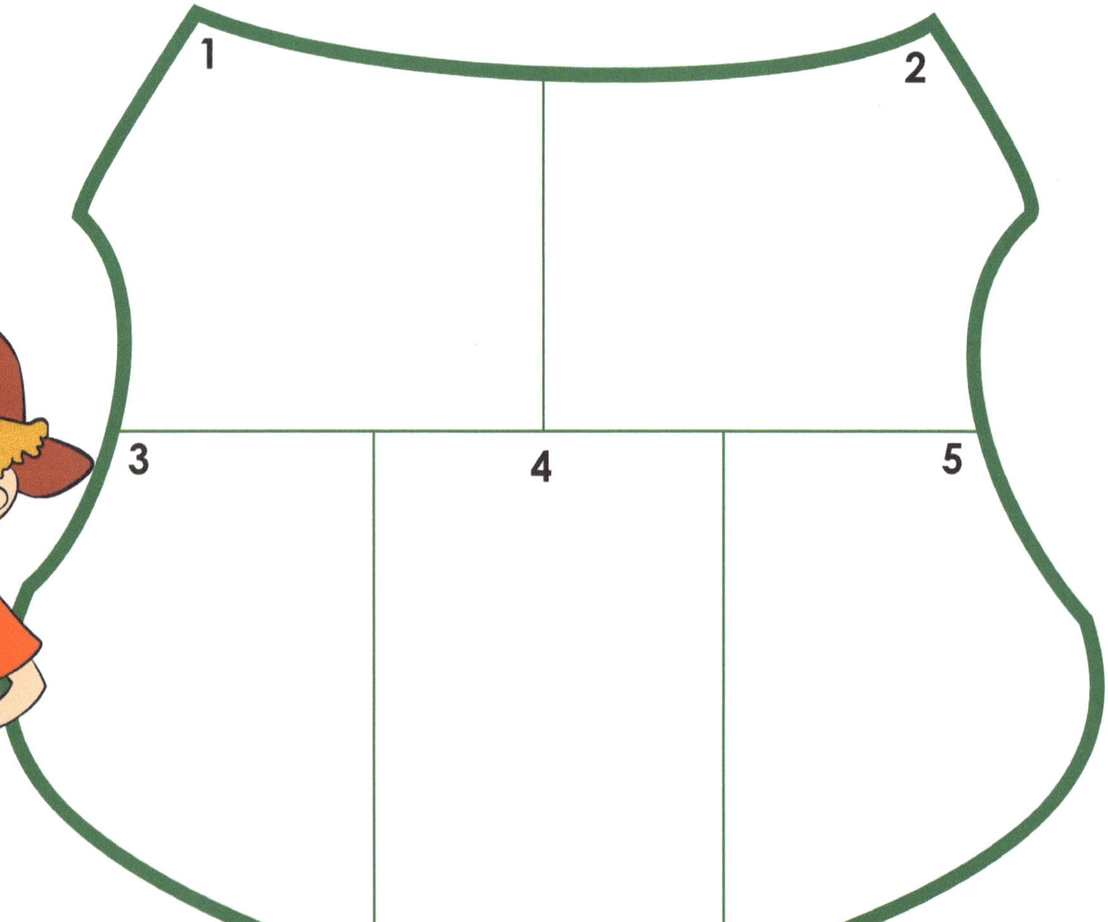

Activity 5
Who Decides?

Your parents and teachers help you make decisions every day. Many important decisions you make for yourself too! You will make more decisions for yourself as you grow up. On this list, circle who decides with you.

Who Makes the Decision?

My favorite book?	Me	Teacher	Parent
What time I go to bed?	Me	Teacher	Parent
If I play a sport in school?	Me	Teacher	Parent
What I eat for dinner?	Me	Teacher	Parent
What I get for a spelling grade?	Me	Teacher	Parent
Where I sit on the bus?	Me	Teacher	Parent
Which friends I have?	Me	Teacher	Parent
What my favorite game is?	Me	Teacher	Parent
If I should help a friend?	Me	Teacher	Parent
When I go to school?	Me	Teacher	Parent
When I take medicines?	Me	Teacher	Parent
If I'm nice to a new classmate?	Me	Teacher	Parent
Where I sit in the classroom?	Me	Teacher	Parent

Copyright © 2017 Dara Wisdom and Empowerment Coaching / Reea Rodney

Activity 6
A Letter to Myself

Have the child write a letter of encouragement to read when they are stressed out. It may feel strange to them as they would be writing about who they are and what they can do, etc. You can put that letter in "My Box of Awesomeness."

Dear _____

Activity 7
Word Finder

Let's see if you can find all the words that describe how confident you are. Search for the words listed beside the puzzle, and circle the words you find. Have fun!

```
C O U R A G E O U S F Q R F U N L R N T
P D B P S D B T J T D F R I E N D L Y U
Y C O L O R F U L R E S P E C T F U L A
V O L C J K L M L O P Y K I T Q L F I Q
K N D A R I N G N Y V M G U B I A A B
M F B P E F D B F G I F T E D X A B B R
I I X A L K S G K H L I L J O Y F U L I
G D K B I C B I F A I E I T C X U L E L
H E V L A L M X V P A R A U P D B O H L
T N Z E B R A V E P G C A R I N G U J I
Z T T Q L F F P W Y Y E Z R X U T S D A
G V H R E N C O U R A G I N G R Q W G N
Y H O N E S T X A S V P O W E R F U L T
```

Capable
Caring
Colorful
Confident
Courageous
Bold
Brave
Brilliant
Daring
Encouraging
Fabulous
Fierce
Friendly
Fun
Gifted
Honest
Happy
Humorous
Joyful
Likable
Mighty
Powerful
Reliable
Respectful
Strong

Activity 8
Fill In the BLANKS

Form a word that best describe who you are; use each letter below which forms the word **CONFIDENCE**. Then give a short explanation or example of how that word describes you.

For example: I am Confident ---- I'm never afraid to answer a question during quiz time.

I am **C** _____

I am **O** _____

I am **N** _____

I am **F** _____

I am **I** _____

I am **D** _____

I am **E** _____

I am **N** _____

I am **C** _____

I am **E** _____

Activity 9
Expressions All About Me

This exercise will allow you to identify all the things about you that makes up who you are. By completing this exercise, you would be able to have a clear picture about your strengths, weakness and goals, so you can build upon it.

I worry about _____

I dreamt about _____

I am proud of _____

I am interested in _____

I am afraid of _____

I don't believe in _____

I am good at _____

I am poor at _____

I have to _____

I feel like _____

I regularly _____

I never _____

I can't stand _____

I have difficulty _____

I have a habit of _____

I no longer _____

_____ makes me laugh.

_____ makes me sad.

_____ makes me angry.

Activity 10
Practice What You're Good At

Write down three talents, skills, or good qualities you do well and enjoy using. Do you make time for each one? Some skills or qualities might be things you can practice every day, like painting. Others might happen once a week — like youth yoga.

Now pick one strength or skill to focus on. **Write down ideas on how you can find time to practice and do it once a week or more. Be specific as to where, when and how often you would be working on this skill or talent.**

For example: I'm a good dancer, and I'm great at taking instructions.
To make this talent a part of my life, I will join the dance group at church or school etc.

1
My Skill, Talent or good quality:

Goal for this Skill, Talent or good quality:

How often will I practice?

2
My Skill, Talent or good quality:

Goal for this Skill, Talent or good quality:

How often will I practice?

3
My Skill, Talent or good quality:

Goal for this Skill, Talent or good quality:

How often will I practice?

Activity 11
Compliment Yourself - Go for it!

Giving a genuine compliment to yourself is an amazing positivity booster! When you take the time to say something nice and give thoughtful praises to yourself, it amplifies your self-confidence and nourishes your self-esteem.

Let's be honest... You rock. There's a lot about you to praise. Building your confidence means recognizing what you are great at. List at least five (5) things that you can praise yourself about. If you are struggling with this, try to ask yourself, "What would a friend say about me?"

1. _____
_____

2. _____
_____

3. _____
_____

4. _____
_____

5. _____
_____

Activity 12
Change the Negative Talk into Positive Talk

The way you speak about yourself reflects the level of self-confidence you have. Switch out the Negative Self-Talk to Positive Self-Talk.

I am ugly

I am not smart

I am a coward

I am a looser

I am not good enough

I am weak

I can't do it

Want more great reading?
Check out these books in our series!

Juniper and Rose

"Check out our Dara Publishing Store at www.darapublishingstore.co for our children's books, clothing, and much more.."

Strictly Essentials Styles by DARA

Visit our website for more: https:www.darapublishing.co/strictly-essentials/